Juan Daniel's *Fútbol* Frog

A Bioengineering Story

Written by the Engineering is Elementary Team
Illustrated by Keith Favazza

Chapter One | Game Day

    Juan Daniel tossed a water bottle into his duffel bag and scanned the rest of its contents. Gonzalez T-shirt? Check. Towel? Check. He was ready to go. Juan Daniel threw the bag over his shoulder and ran through the kitchen towards the front room of the house. Friends and neighbors were sitting at small tables eating Mamá Tere's delicious food. Mamá Tere's *pupusería* was always busy on weekend afternoons. Just as Juan Daniel stepped outside into the hot Salvadoran sunlight, ready to sprint to the field, a voice called to him.

    "¡Juan Daniel! Not even a kiss for your grandmother?"

    It was Mamá Tere. Juan Daniel skidded to a stop, ran back inside, and placed a kiss on her cheek. "I'll be home

after the game!" he called, already out the door.

"¡*Buena suerte*!" Mamá Tere called. She shook her head and smiled before turning back to her customers.

Chapter Two | Benched

As Juan Daniel tore down the dusty road, he looked toward the thick trees in the distance. Although Juan Daniel's town was located on the edge of *El Imposible*, a rain forest reserve in El Salvador, he knew from experience that *el campo de fútbol*—the soccer field—would be hot and dry. There were no trees to block the beating sun, and since it was the middle of the dry season, it hadn't rained for months.

"*¡Hola*, Carlos!" Juan Daniel yelled to his best friend. Carlos waved to him and gave the ball a fierce kick

halfway across the field. Juan Daniel jogged out to him and started warming up.

"Hey, nice shirt! Is it new?" Carlos teased. Juan Daniel wore this T-shirt to every soccer game. He'd gotten it when he had visited the capital, San Salvador, two years ago. While

he was there he had seen his soccer idol, Jorge *"El Mágico"* Gonzalez—number 11 on the national team—in a parade. This T-shirt was a replica of *El Mágico*'s soccer jersey.

"Are you ready for the big game?" Carlos asked.

"More than ready," Juan Daniel said. Juan Daniel and his friends had been playing soccer against other boys from their town every weekend for almost a month. There were two games left to determine the neighborhood champion. Juan Daniel knew his team could win—they just had to focus.

"You guys ready?" called José Eduardo, the captain of the other team. Juan Daniel looked up at him and nodded. José Eduardo towered over all of the other boys on the field. *He may have size*, thought Juan Daniel, *but I have speed!*

From the game's start, Juan Daniel proved himself right by quickly running up the field to score the first goal. The teams lined up for the kickoff, and play started again. Juan Daniel got control of the ball and sprinted up the field. He saw the goalie hunkering down, getting ready to block his shot. Juan Daniel brought his foot back. Just as he was about to connect with the ball, he felt a hard push against his shoulder. It was José Eduardo.

Juan Daniel hit the ground. Carlos jogged over and knelt down by Juan Daniel. "Hey, he's just a bully. Forget him," Carlos said. Juan Daniel wiped the dust out of his eyes and moved to hoist himself up.

"Ahhh!" Juan Daniel cried out as he leaned on his arm. He crumpled. "My arm," he gasped. More of Juan Daniel's teammates ran over as they saw him hit the ground again.

"Faker! We know you're just playing," called one of José Eduardo's teammates.

"What are we going to do?" asked Carlos.

"I can play, I swear," said Juan Daniel. "I just fell on it hard. It won't even bother me."

"No, Juan Daniel," Javier, the team's forward, broke in. "This is a really important game, but we need you to be able to play in the next game, too. What if you hurt your arm even more by swinging it as you run? You should at least rest for a little bit."

Juan Daniel saw ten faces nodding in agreement. With his good arm, Juan Daniel pushed himself up and headed toward the bench.

Chapter Three | A *Fútbol* Frog

Juan Daniel sat out the rest of the first half. His teammates were holding their own, but it was a close game. As he stared out onto the field, something popped into view. It was a frog. Juan Daniel watched it hop around the edge of the field. He scooped up the frog and looked into its bright, gold eyes. Running a finger along the frog's green and brown patterned skin, Juan Daniel paused. "Hey, it seems like there's something wrong with you," he said. The frog's skin was dry like paper—nothing at all like the moist skin of frogs he'd found in the rain forest.

"What's that?" asked Carlos after slugging down a mouthful from his water bottle. Juan Daniel jumped in surprise. He had been so interested in the frog that he hadn't noticed that Carlos had taken a water break.

"I found this frog on the edge of the field," Juan Daniel said. "Don't you think it's strange that he's out here? Usually I find frogs in *El Imposible*, hanging out under rain forest plants and logs where it's wet and cool. This frog is out here in the blazing sun."

"The other day my papá was talking about an article he read in the newspaper," Carlos said, still catching his breath. "With so much of the rain forest being cut down, a lot of animal habitats are being lost. The animals have nowhere to go and sometimes end up in the wrong place." Carlos looked down at Juan Daniel's arm. "How are you feeling?"

"My arm feels a lot better," Juan Daniel said. And it did feel much better, but Carlos gave Juan Daniel a look of disbelief before he jogged back onto the field.

"Looks like we've got something in common," Juan Daniel said to the frog. "We're both in the wrong place here on the sideline."

Juan Daniel grabbed his water bottle and carefully sprinkled water on the frog. "Maybe that'll make you more comfortable until the end of the game," he said.

As the frog settled into the shade of Juan Daniel's shadow, Juan Daniel watched his teammates. With just a few minutes left, Juan Daniel knew his team would need some

quick moves to get a goal. Juan Daniel noticed his teammate Ernesto looked tired. Just then, Ernesto jogged toward the bench and reached for his water bottle.

Juan Daniel began talking a mile a minute. "Ernesto, you've got to let me in for the last minute. I know I can do this. My arm's feeling a lot better and all I have to do is sprint down the field and get one good kick in—"

Ernesto was huffing and puffing, taking sips of water between breaths. "Does your arm really feel better?" he asked.

"It really does, I promise," Juan Daniel pleaded.

"Okay," Ernesto said slowly. "Go get 'em."

A huge grin spread across Juan Daniel's face. He tightened his shoelaces and reached for his water bottle to take a drink. Sitting right next to it, as if waiting to wish him luck, was the frog.

"Watch this," Juan Daniel said to the frog as he ran to get into position.

Juan Daniel took off toward midfield. With adrenaline racing through his veins, he barely felt a thing as he pumped his arms while he ran. He glanced back over his shoulder and saw Carlos kick the ball through the air in a high arc. As soon as it landed in front of him, Juan Daniel began dribbling the ball down the field. Nearing the goal, Juan Daniel glanced to each side for defenders, then booted the ball hard toward the goal. It seemed to move in slow motion as it sailed through the goal posts.

"Woo-hoo!" yelled Carlos, giving Juan Daniel a congratulatory slap on the back. All of his teammates soon crowded around.

"That was awesome," said Mario. "How'd you manage

to come back so strong after that fall?"

Juan Daniel shrugged. "I don't know—I just knew I could do it. Plus, I had a little luck from that frog," he said, pointing to the frog, still sitting near his water bottle.

"A frog?" asked Mario.

"Yeah, a frog. But not just any frog. A lucky frog. And our new team mascot," said Juan Daniel. "I think everyone—including the frog—should head back to my grandmother's *pupusería*. We need some victory food."

Chapter Four | A Helpful Visit

Later that day, after the celebratory dinner with his team, Juan Daniel helped Mamá Tere serve plates of steaming *pupusas*. On a table in the corner, the frog looked on from a large bowl that Juan Daniel had placed him in.

"Juan Daniel," Mamá Tere called. "Can you help those people who just came in?"

I wonder what they're doing here, Juan Daniel thought. It wasn't often that outsiders stopped in Juan Daniel's town. They usually just passed through on their way to the rain forest. There was a blond woman with an accent and a Salvadoran man and woman, both in business suits. Juan Daniel walked over to take their order.

"I think that with one more trip into the rain forest,

A Helpful Visit

we'll have the data we need to go back to the lab," said the Salvadoran man.

"Well, that's only if the frogs don't hide themselves away in a bunch of leaves!" added the blond woman.

"Frogs!" Juan Daniel cried out before he even realized he'd opened his mouth. "I found a frog today!" Juan Daniel pointed towards the corner table where he had left the frog.

"It looks like frogs are everywhere you least expect them," laughed the blond woman as she stuck out her hand. "Hi, I'm Kristin Peters. My colleagues and I have been studying frogs in *El Imposible*." Juan Daniel introduced himself and shook her hand. "Where did you find that frog?" she asked.

Juan Daniel explained to Ms. Peters about the dusty soccer field and how the frog had been his team's lucky mascot.

"I'm glad he brought you good luck," Ms. Peters said. "These frogs are having an awfully hard time with so much of the rain forests being cut down. But it looks like you're taking pretty good care of him. Just make sure to keep him moist. Frogs need moist skin so they can absorb air into their bodies."

"I poured some water on him at the game," Juan Daniel said. "But hopefully I'll play in all of the next game, so I won't be around to pour water on him. I'll have to think of a better way to keep him moist."

Juan Daniel felt a hand on his shoulder and turned to see Mamá Tere smiling and shaking her head. "Juan Daniel, I think these people stopped by for food, not to have you talk their ears off."

"It's OK," said Ms. Peters. "He was just telling me about his frog. Juan Daniel, if you have a few minutes, you might be interested to hear about some of the work that I do. I bet it would help you come up with a way to help your frog."

Juan Daniel looked hopefully at Mamá Tere. "Well, I can't have that frog sitting in the *pupusería* forever, can I?" Mamá Tere asked. "Go ahead and sit and figure out what to do with it."

Juan Daniel grinned and pulled up a chair. Ms. Peters explained that she traveled into the forest in search of amphibians, such as frogs, so she could study the special properties of their skin.

"Frog skin is pretty neat," Ms. Peters said. "Frog skin—and people skin—is a membrane. Membranes protect us like a shield, keeping harmful things out of our bodies. But they also let some things pass through, like water and oxygen. That's how frogs drink—they absorb water through their skin. Some frogs have skin with what we call anti-microbial properties. Their skin can fight bacteria and viruses. Can you imagine if people could come up with some sort of coating like that for our skin? It could stop people from getting sick, or even help us come up with new medicines or vaccines."

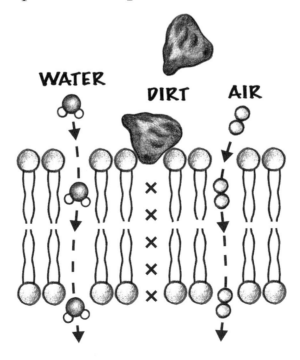

Juan Daniel nodded. "That's cool. So are you a scientist?"

Ms. Peters shook her head no. "The two people I'm with are scientists. It's really important to work closely with scientists in my job, though. I'm a bioengineer. I use what I know about math and science—especially biology, or things found in nature—to solve problems that living things might have. Like people, or your frog."

"I guess figuring out how to keep my frog's skin moist during the game is an engineering problem," said Juan Daniel.

"It sure is," said Ms. Peters. "How were you thinking of solving it?"

"I haven't thought too much about it yet," said Juan Daniel.

"I have an idea that might get you started," said Ms. Peters. "You could try taking a walk outside. Sometimes nature has already solved a problem in a unique way. When bioengineers look at how nature works, we can get some great information that helps us create technologies—things or processes that help us solve problems."

"So I could create a technology that helps solve my problem," said Juan Daniel. "That's a great idea. I think I'll take a trip to the rain forest."

By now the two Salvadorans were done with their *pupusas*. As they slid their chairs back from the table, Juan Daniel stood to say goodbye.

"I tell you what," said Ms. Peters. "I'll be back here in a few days. Maybe I can stop by to see how your solution is coming along."

As Ms. Peters and her coworkers walked out the door, Juan Daniel sat back down at the table. *Who knew that there were engineers out there studying and trying to design something like frog skin?* he thought. *And who knew they traveled to places like my little town in El Salvador?*

Chapter Five | A Trip to the Rain Forest

Before the sun sank below the horizon, Juan Daniel walked over to the edge of the forest in his town that led to *El Imposible*. He stood in a small clearing and looked straight up into the leafy canopy above him. Faint rays of sun filtered through and made a spotty pattern on the ground.

Juan Daniel tilted his head to listen to a blue-crowned *motmot*—the national bird of El Salvador—that was flying nearby. He caught a glimpse of the bird's orange throat and blue wings before it disappeared into the forest.

Juan Daniel heard a rushing sound and headed out of the clearing toward it. As he got closer, he realized what it was: a waterfall. As the sheet of cascading water came into view, Juan Daniel grimaced. He was imagining his poor frog sitting underneath the waterfall. *I don't think he'd stand a chance being hit with all of that water*, Juan Daniel thought.

He saw the trees and bushes at the base of the waterfall and thought of what Ms. Peters had said about the frogs hiding themselves inside leaves. He walked towards the tree and rubbed the waxy green surface of a leaf. Suddenly he noticed a sparkle of light out of the corner of his eye. He looked closer.

The light reflecting off of a bead of water on one of the leaves made the drop shine like a tiny jewel. Juan Daniel watched the water slowly slide down the leaf and make a tiny splash. He reached out to catch a drop in his hand.

That's it! Juan Daniel realized. *I need to make something that will drip just enough water onto the frog to keep him moist. Now I need to figure out how to design it!*

A Trip to the Rain Forest

Chapter Six | Modeling a Membrane

The next day after lunch, Juan Daniel sat inside the *pupusería* wondering how he could design something to help his frog. His sister Marcela walked in and noticed him looking intently into the frog's bowl.

"You're still thinking about that frog?" she asked.

"Well, he's a champion, just like I want to be," Juan Daniel answered. "Here it is, the middle of the dry season, and he was out in the bright hot sun, away from his natural home, but he was hopping right along. Before I bring him back to the rain forest, I want him with me at our last game."

"You'll win again. I'm sure of it!" said Marcela.

Juan Daniel smiled at her. "I hope so. But there is one more thing that I have to figure out before this frog can help lead us to victory." Marcela raised her eyebrows in a question. "This frog is supposed to live in the rain forest, where it's moist. Out in the field it's hot and dry."

"So what are you going to do? Give him a teeny tiny glass of water?" Marcela giggled.

"No, frogs don't drink water like we do. I learned the other day that they absorb it through their skin. But I bet I can come up with a way to get water to him," Juan Daniel said. "I've already started imagining some ideas."

"You sound like a scientist," said Marcela.

"No, like an engineer," Juan Daniel said. "They use science to help them design solutions to problems, just like this one." Juan Daniel began telling Marcela about his talk with Ms. Peters and his discoveries in *El Imposible*. Then he grabbed a piece of paper and started sketching out a bunch of the ideas he'd imagined. "What I think we need," he said as he continued sketching, "is a type of membrane."

"A membrane?" Marcela asked.

"Well, a model of a membrane. It will separate the container where the frog lives from a container filled with water. The membrane will let some water drip through—just enough to keep the frog moist."

"I get it," said Marcela. "It'll be as if the frog lived under our water tank and we poked a couple holes in it so

he'd get sprinkled with water."

"Yes!" cried Juan Daniel. "That's exactly the type of thing that we need to do. Only we have to make sure he's really just showered, not flooded. If there's too much, I bet the water would be gross after a few days of the frog living in it. And we can't change the water too often, since we only get water from the well twice a week."

"It sounds like you're off to a great start!" Juan Daniel turned and saw Ms. Peters standing behind him.

"Oh, great, you're back!" Juan Daniel said. He introduced his sister, then turned back to the pictures he had drawn. "I have some ideas, Ms. Peters," Juan Daniel said. "But I'm not exactly sure what to do next."

"Maybe you'll want to use the steps that I find helpful in my work. They're called the engineering design process. First you ask some good questions and imagine lots of possible solutions, just like you've started to do already. Then you think about what materials you could use, make a plan, and create your design. Finally, you see if there's any way to improve it."

Juan Daniel looked from Ms. Peters to his lucky frog. "That sounds like a good idea," Juan Daniel said. "I think I can do that."

"I'll help you get started," said Marcela. "We could test some materials that we have around the house to see how well water moves through them. What about one of the sponges from Mamá Tere's kitchen?"

"Yeah, and I have an old T-shirt that we could cut up," said Juan Daniel.

"That's great," said Ms. Peters. "I know you two will come up with something wonderful. I wish I could see it, but I came to say goodbye and wish you luck. I leave for the university tomorrow, and I won't be back for a long time."

Modeling a Membrane

"You're leaving?" Juan Daniel asked. His eyes darted around the *pupusería*. It was a Salvadoran custom to give a guest something before they left. His eyes lit up as he spotted the dried gourd hanging from a colorful string on the wall. He pulled it down and handed it to Ms. Peters.

"This is a *tecomate*," he said. "In El Salvador, we use these to collect water. I know it's not the best way to get water to my frog, but maybe you will remember me by it."

"Thank you, Juan Daniel. I will." With a smile and a wave, Ms. Peters was out the door.

Together, Juan Daniel and Marcela tested sponges, fabric, a plastic bag with holes punched into it, and some other materials that they found around the house. They tried putting some of the materials together to see how water traveled through both of them. When they thought they had figured out a combination of materials that worked well, they decided to leave them set up overnight to see how much water had traveled through their model membrane by the next morning.

After a few improvements, Juan Daniel thought his model membrane and the frog would be ready for the big game. *But the question is*, thought Juan Daniel, *am I ready?*

Chapter Seven | Champions?

On the morning of the big game, Juan Daniel woke with the sun. He peered out the window and saw peach rays of light reaching across the sky. He could feel the fluttering of nervousness in his stomach, but also excitement and determination in his heart.

Juan Daniel wandered outside and started running drills. All he could think about was *fútbol*. Before he knew it, it was time for the game. He picked up the frog, sitting in a bowl under the membrane he had designed, and headed to the field.

Juan Daniel's team kept the game tied until the end of the first half. Then, two well-placed kicks from José Eduardo's team left Juan Daniel's team two goals behind.

As halftime began, the team gathered at the bench. Juan Daniel glanced toward the frog, who was still hopping around, cool and comfortable, under his newly-designed membrane. *We need some luck now more than ever*, thought Juan Daniel.

Carlos's voice rose in a panic. "What are we going to do? In just five minutes they've scored two goals!" he squeaked.

"Guys!" cried Juan Daniel. "We can do this—I know it!" Juan Daniel paused for a few moments as an idea clicked in his mind. "You know, I learned about this thing the other day—this process. Maybe it could help us here. We need to ask, imagine, and plan."

"What are you talking about?" asked Carlos.

"We're gonna win this game! I'm talking about how we'll do it—not just by using our legs and feet, but also by using our minds to outsmart the other team. We need to ask some good questions. What patterns have these guys been running all day?"

"Up the right side to Angelo, their fastest forward," answered Javier.

"Yeah, and every time we're too late to block the pass. We need to imagine some ways to defend our goal, and then come up with a plan. Is everybody with me?" Juan Daniel placed his right hand in the center of the circle of teammates. Carlos stuck his hand on top of Juan Daniel's, and Javier, Cesar, and the rest of the team followed.

After they'd come up with a solid plan, Juan Daniel's team ran onto the field with newfound determination. José Eduardo's team looked shocked as Juan Daniel's team scored a goal within their first three minutes back on the field. Then they scored another to tie the game. But with one minute left,

Juan Daniel found himself still eight meters away from the goal, with José Eduardo guarding him like a hawk.

Juan Daniel got control of the ball and took off running. José Eduardo was close on his heels. Juan Daniel felt a heavy hand on his shoulder, just like last week when he was knocked down and hurt. This time he guarded the ball closely between his feet, ducked down to get out of José

Eduardo's way, and continued to run up the field. Without Juan Daniel there to push on, José Eduardo lost his balance and went flying toward the ground. He landed just in time for the dust from Juan Daniel's shoes to billow up in his face. Juan Daniel shot the ball forward. Goal!

Juan Daniel's teammates ran to him, whooping and cheering.

"That shot was amazing!" cried Carlos. "The plan really worked! I can't believe we won."

Juan Daniel smiled at him. "There's just one more thing that I need to take care of. Meet me at school early tomorrow morning, okay?"

Juan Daniel grabbed his bag and the frog and headed off toward home.

Chapter Eight | Freeing the Frog

The next morning Juan Daniel sat outside his school, with the membrane he had designed and the frog. On a piece of paper he carefully drew a diagram of his frog container and membrane, placed it in an envelope, and addressed it:

Ms. Kristin Peters
Facultad de Ingeniería y Arquitectura
Universidad de El Salvador
San Salvador, El Salvador

As he sealed the envelope, he saw Carlos walking toward him.
"All right, I'm here," Carlos said. "What did you want to take care of?"

"It's not for me," said Juan Daniel. "It's for our lucky team mascot!" Juan Daniel led Carlos to the back of the school. Leafy canopy stretched above them. Juan Daniel carefully loosened the membrane from the top of the frog's bowl and watched with approval as the frog hopped out and headed towards a low, leafy bush nearby.

"You didn't want to keep him? Don't you think we need him to be our mascot in the tournament next time?" Carlos asked.

"Nah," Juan Daniel said. "He definitely inspired me, but next time I think we'll do fine with our determination, our *fútbol* skills, and a great plan."

Freeing the Frog

Try It!

Design a Model Membrane

You can design a model membrane, just like Juan Daniel! Frogs need to absorb water through their skin, which is a kind of membrane. Imagine that you have a pet frog and you need to find a way to make sure it gets enough water. Your goal is to design a model membrane that slowly drips water into your frog's habitat.

Materials
- ☐ Clear plastic cup
- ☐ Small plastic bowl
- ☐ Piece of wire mesh
- ☐ Rubber bands
- ☐ Tape
- ☐ Scissors
- ☐ Sponges
- ☐ Plastic wrap
- ☐ Aluminum foil
- ☐ Cotton balls

Setup

Ask an adult to help you cut a hole in the bottom of the clear plastic cup. This cup will hold the water that will drip through the model membrane you design.

Make a model frog by cutting a piece of dry sponge into a frog shape. Place it inside the plastic bowl. Put a piece of wire mesh on top of the bowl and secure it in place with a rubber band.

Design Your Model Membrane

Water must move through your model membrane slowly. Think about the materials you have. Are any of the materials waterproof? Can you change any of the waterproof materials so that they allow just some water to pass through? Should you poke holes through any of the materials? How many holes?

Secure the model membrane that you design over the top of the plastic cup using a rubber band.

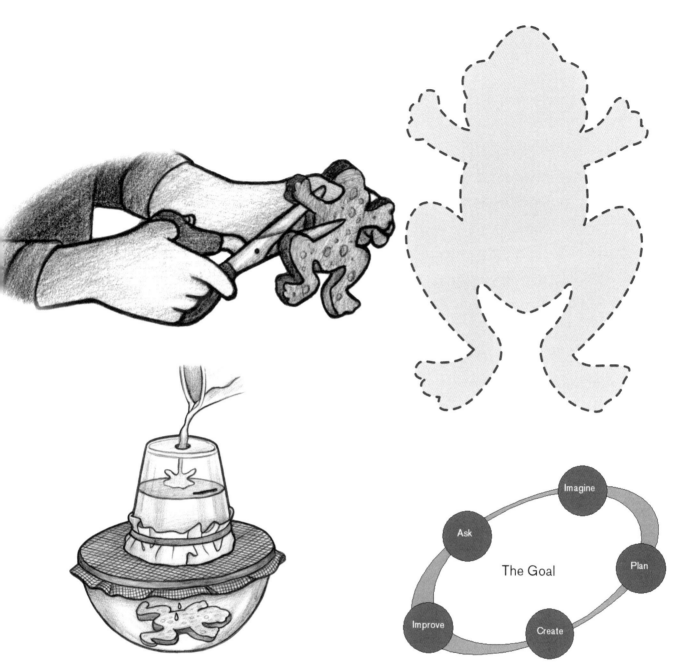

Test Your Model Membrane
Turn the plastic cup upside-down and place it on top of the wire mesh you secured to the bowl. Pour about 1/2 cup of water through the hole in the bottom of the plastic cup. Using a marker, draw a line at the top of the water in the cup. How long does it take for the water to move through your model membrane and into the frog container? A few minutes? A few hours? A day? Keep checking on your design and mark the water level on the plastic cup each time you observe it.

Improve Your Model Membrane
Did water move through your model membrane too quickly? Too slowly? Not at all? Use the engineering design process to improve your model membrane.

See What Others Have Done
See what other kids have done at http://www.mos.org/eie/tryit. What did you try? You can submit your solutions and pictures to our website, and maybe we'll post your submission!

Glossary

Adrenaline: A chemical that the body releases into the bloodstream in response to stress, fear, or excitement.

Amphibian: A cold-blooded animal with a backbone capable of living on both land and water. Examples include salamanders and frogs.

Anti-microbial: Able to stop or impede the growth of tiny disease-causing organisms.

Bioengineer: Someone who combines his or her knowledge of science, math, and living things to design technologies that solve problems in nature or use natural materials to solve man-made problems.

Buena suerte: Spanish phrase for good luck. Pronounced *bwen-ah swer-tay*.

Campo de fútbol: Spanish phrase for soccer field. Pronounced *camp-oh deh fooht-boll*.

Canopy: The highest layer of leaves and branches in a forest.

El Imposible: A rain forest wildlife refuge in El Salvador.

Engineer: A person who uses his or her creativity and understanding of mathematics and science to design things that solve problems.

Engineering Design Process: The steps that engineers use to design something to solve a problem.

Fútbol: Spanish word for soccer. Pronounced *fooht-boll*.

Habitat: The environment where an animal or plant normally lives.

Hola: Spanish word for hello. Pronounced *o-lah*.

Membrane: A very thin structure through which substances can move based on their size, shape, concentration, and electric charge, as well as other factors.

Pupusa: A cornmeal cake that can be stuffed with cheese, refried beans, pork, or chicken. Pronounced *poo-poo-sah*.

Pupusería: A restaurant in which *pupusas* are served. Pronounced *poo-poo-seh-ree-ah*.

Rain forest: A dense, tropical forest with more than eight feet of rain per year.

Technology: Any thing or process that people create and use to solve a problem.

Tecomate: A dried gourd used for holding water. Pronounced *tay-coh-mah-tay*.